by Janette Schuster

Printed in Mexico

ISBN 978-0-15-362393-6
ISBN 0-15-362393-4

2 3 4 5 6 7 8 9 10   126 16 15 14 13 12 11 10 09 08

Harcourt
SCHOOL PUBLISHERS

Visit *The Learning Site!*
www.harcourtschool.com

Introduction

What do you think of when you hear the word <u>storm</u>? Do you think of booms of thunder? Do you think of pouring rain? Storms can be dangerous. Knowing about the dangers can help you stay safe.

Thunder and Lightning

A thunderstorm has strong winds and heavy rain. It has lightning and, of course, thunder. Sometimes so much rain falls that it causes floods. Sometimes a storm has a lot of lightning. Lightning can hurt or kill people.

lightning

No matter where you live, you have probably seen a thunderstorm. They happen in most places on Earth. In fact, right now, about 2,000 thunderstorms are going on in places around the world!

Thunderstorm Safety

1. What to Do ☺
   - Go inside a strong building right away.
   - If you can not get into a building, get into a car or crouch down as low as you can.

2. What Not to Do 🚫
   - Do not go under a tree.
   - Do not use electric appliances.
   - Do not talk on a phone that has a cord.
   - Do not take a shower or a bath.
   - Do not touch metal objects such as fences or flagpoles.

Twisting Air

Have you ever seen a movie of a tornado? If so, you know how scary a tornado can be!

A tornado can come out of a very strong thunderstorm. A tornado forms between the clouds and the ground. It has very strong winds. It can lift up cars and knock down houses.

tornado

When a tornado forms, meteorologists send out a warning by TV or radio. A meteorologist is a scientist who studies weather.

The United States gets about 1,000 tornadoes every year. That's more than any other place on Earth!

Tornado Safety

1. What to Do ☺
   - Go inside a strong building right away.
   - Go to the basement. If there is no basement, go to the lowest floor. Stay in a closet or hallway. Stay away from outside walls and windows.
   - If you can not get inside, lie flat in a ditch or low area.

2. What Not to Do ⊘
   - Do not stay in a car or a moblie home. Go inside a strong building instead.
   - Do not stay in a large, open room.
   - Do not open windows. Take cover instead.
   - Do not go near windows and doors.

Name that Storm

What kind of storm can you call Hugo or Hazel? A hurricane! To help keep track of hurricanes, meteorologists give them names. Hurricanes are huge storms. A hurricane can be more than 300 miles wide. It has very strong winds and brings heavy rain.

hurricane

Large, slow hurricanes take a long time to pass. This means that a hurricane can rain on one place for many hours. So much rain can cause floods.

A hurricane's strong winds can push ocean waves onto land. This also causes floods. The strong winds and floods can destroy homes and kill people.

Hurricane Safety

1. What to Do 😊
   - Go as far away from the ocean as you can.
   - Go indoors to a basement, closet, or hallway. Stay away from windows and doors.
   - Listen to a portable radio or TV. Do what the meteorologists tell you to do.
   - If you are told to leave your home, go to a strong building or a shelter. Tell someone outside the storm area where you are going.

2. What Not to Do 🚫
   - Do not stay in a mobile home. Go inside a strong building.
   - Do not go near flooded places.
   - Do not go outside until it is safe.

Take Action Now

The best way to stay safe is to plan before a storm comes. Talk to your friends. Talk to your family. Tell them what you have learned.

Before a hurricane comes, plan how you and your family will leave the area. Plan where you will go. Have supplies ready. Make sure that you have flashlights, extra batteries, a radio, a first aid kit, food, and water.

Be prepared!